Drilling at Wheal Jane Tin Mine, near Truro, Cornwall.

TIN AND TIN MINING

R. L. Atkinson

Shire Publications Ltd

CONTENTS

Set in 9 point Times roman and printed in Great Britain by C. I. Thomas & Sons (Haverfordwest) Ltd, Press Buildings, Merlins Bridge, Haverfordwest, Dyfed.

COVER: *West Wheal Owles 36 inch Pumping Engine House, Cargodna Section.*

ACKNOWLEDGEMENTS

I am indebted to the following colleagues at the Camborne School of Mines for their help with technical data: Dr R. Barley, Dr A. V. Bromley, Mr J. Shrimpton, Mr J. Turner, Dr B. Wills and the AVA department for printing the photographs. Thanks are also due to Mr R. Penhallurick of the Royal Institution of Cornwall for his assistance. Special thanks are due to my husband, Keith, and daughters for their help and encouragement. Photographs on the following pages are acknowledged to: H. Boerner (Royal Institution of Cornwall) 22 (lower); Dr A. V. Bromley 3, 4, 5, 6 (upper), 8 (left), 9 (upper), 10 (lower), 12, 13 (upper, left), 29; J. C. Burrows (Royal Institution of Cornwall) 13 (lower), 14, 15, 16, 17, 18, 23, 24 (upper); Carnon Consolidated 1, 24 (lower), 31; Carnon Consolidated and J. Watton 2, 28; Tony Clarke (Camborne School of Mines) cover; Professor K. Hosking 25; H. W. Hughes (Royal Institution of Cornwall) 27 (upper); Royal Institution of Cornwall 6 (lower), 7, 8 (right), 9 (lower), 10 (upper), 13 (upper, right), 20 (lower), 21, 22 (upper), 26 (lower); South Crofty Mines Ltd 27 (lower); J. Trounson 21 (lower); J. Trounson and S. J. Govier 22 (upper); Dr B. Wills 19.

A ball mill (foreground) and rod mill (background) at Wheal Jane, 1975. Crushed ore is introduced into the rotating drums and ground between metal balls or rods inside them.

A tin lode about 1 inch (25 mm) wide in Cornish granite.

GEOLOGY AND EXPLORATION

Tin was one of the first metals to be used by man but it was not employed on a large scale until the nineteenth century. Cassiterite, the tin ore, occurs in veins or *lodes* formed at high temperatures within the earth's crust. It is usually associated with granites. The lodes vary in width from under an inch (25 mm) to several yards and may be half a mile (800 m) long. Sometimes they contain important concentrations of ores of other metals mixed with unwanted *(gangue)* minerals such as quartz, tourmaline and chlorite.

Cassiterite, tin oxide, is hard, heavy and usually brown or black in colour, although when pure it can be white. The crystals are generally short and square in section but can be long and slender (needle tin) or form dull crusts (wood tin). A very minor ore of tin is stannite, a complex sulphide in which tin occurs with copper, iron and often zinc. Stannite occurs in small amounts in the tin-bearing veins of Cornwall and Bolivia.

Tin is still worked in south-west Eng-land, where there has been a long history of tin mining. During the second half of the nineteenth century Cornwall provided nearly half the world's tin. The rocks which form the peninsula of south-west England are between about 260 and 360 million years old. They accumulated as sands and muds and as the ocean floor slowly subsided a pile of sediment more than 6 miles (10 km) thick built up. Sometimes coral reefs reached up nearly to sea level and great submarine volcanoes erupted lava and volcanic ash.

When Coal Measure swamps covered much of northern England mighty compressional forces crushed the great pile of sediment. This was turned into hard rock and the original layers were thrown into great folds. The whole region was lifted into a great mountain chain perhaps as high as the Himalayas. The lower regions, beneath the mountain chain, began to melt and molten rock (granite magma) rose up as great cylindrical masses and began to crystallise. As crys-

3

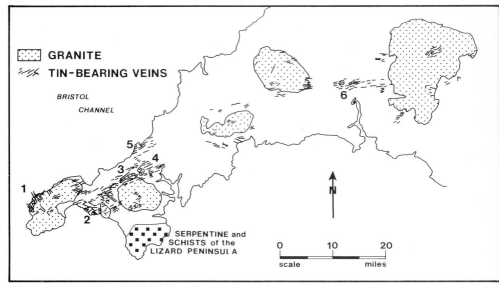

GRANITE

TIN-BEARING VEINS

BRISTOL
CHANNEL

6

1

5
4
3

2

N

SERPENTINE and
SCHISTS of the
LIZARD PENINSULA

0   10   20

scale   miles

Simplifed geological map of south-west England showing the occurrence of tin-bearing veins (lodes) and granite masses. Surrounding the granites, and unornamented on the map, are shales, sandstones, limestones and volcanic rocks known collectively as 'killas'. Principal tin-producing regions are: 1. St Just-Botallack; 2. Mount's Bay; 3. Camborne-Redruth; 4. St Day; 5. St Agnes-Perranporth; 6. Kit Hill-Gunnislake. Tin mines still operate in the St Just, Camborne-Redruth and St Day areas.

tallisation proceeded a watery residue rich in ore metals accumulated near the top of the granite.

As the rock cooled it formed cracks, some only a hairline's thickness, others hundreds of feet deep, sometimes miles long and often more than 3 feet (1 m) across. The watery residue streamed out of the granites along the cracks and as it cooled it deposited a variety of ore minerals. These mineral-filled lodes concentrated around the edges of the granite masses. A long period of erosion wore down the land before the granite was exposed at the surface and the lodes were accessible for exploitation.

During this time the easily decomposed minerals of the sedimentary rocks and the granites were broken down to fine particles of clay, washed into rivers and carried away to the sea. The ores of copper, lead and zinc were slowly dissolved by surface waters and they too were dispersed and lost. Tin ore is very resistant and very heavy and tends to accumulate in potholes, deep channelways and temporary lakes. In such places

it may be concentrated sufficiently to form workable ore deposits. These are purer than the cassiterite from underground lodes and are easier to process.

Early prospectors found these *placer* deposits, then worked upstream until they reached the tin-bearing veins. Modern exploration relies on a systematic study of large regions and aerial (sometimes satellite) photography and remote sensing techniques are employed. In geophysical exploration the physical properties of rocks such as magnetism, density, radioactivity and electrical resistance are measured with special instruments. Geochemical exploration aims at finding unusual concentrations of a metal in a particular area by the chemical analysis of rocks, soils, sands, mud, vegetation and stream water.

If an unusual *(anomalous)* area has been located by geophysical or geochemical survey diamond drilling is undertaken around the suspected ore body to establish its size and shape and to recover samples for testing. Estimates are made of expected tonnage and grade of the ore

Two forms of cassiterite. Brown wood tin from Mexico and black, octahedral crystals from Zimbabwe — the largest is ¾ inch (19 mm) across.

and the amount of waste rock which will have to be removed. Work is carried out in the laboratory, where core is inspected and logged. Pieces are cut and polished for inspection under a microscope. When the tests are complete the geologist can write his report indicating whether the area is suitable for mining.

Diagram showing the formation of placer deposits. Primary tin veins are eroded by fast-flowing rivers in upland regions. The cassiterite liberated from the lodes is carried in suspension until the river slows down or where it enters the sea. There the cassiterite is selectively deposited where it may form secondary alluvial deposits.

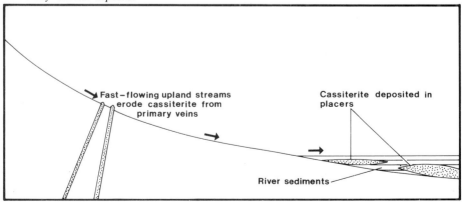

Fast–flowing upland streams erode cassiterite from primary veins

Cassiterite deposited in placers

River sediments

ABOVE: *An X-ray fluorescence spectrometer is designed to determine the types and amounts of metals in a sample by the radiation they emit when bombarded with X-rays.*

BELOW: *This almost pure tin bowl with a lid dates from Roman times and was found in a Cornish tin stream. It is about 18 inches (457 mm) wide and is exhibited at Truro Museum. The function of the bowl is not known but the handles have been mended several times rather crudely so it seems to have been a well-used item.*

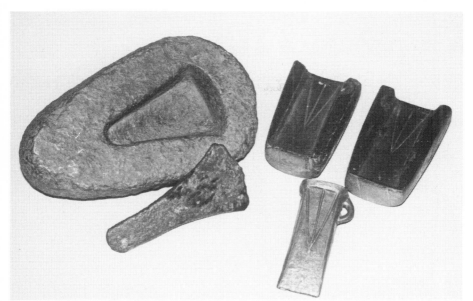

Bronze axe heads and moulds. Early bronze age axe heads are characterised by open moulds whilst late bronze age artefacts were frequently cast in double moulds and show less sign of hammering.

HISTORY

Although tin was one of the earliest metals known there is little written information about its discovery and use. There is a reference to tin in the Old Testament and it is mentioned in classical literature by, amongst others, Herodotus (about 450 BC) and Pliny (who died in AD 79). It has been proved that Britain was trading in tin three thousand years ago but a widespread belief that the Phoenicians traded in Cornish tin has been proved erroneous.

Very few pure tin objects have been found from antiquity. This may be because tin, being a rare metal, was too valuable to use other than as thin sheets for decoration, or the tin may have disintegrated. It could have been that tin artefacts either never existed or were melted down at a later period for more useful alloys. At least two areas with well known tin deposits, Cornwall and Saxony, also yield copper so the two metals may have been combined accidentally to form one of the most useful early alloys, bronze. This metal was so important that

it gave its name to an entire period of human history, the bronze age.

In the Near East copper metal was used as far back as 7000 BC. The addition of tin for making bronze is seen intermittently from 3000 BC onwards but was not common until about 2000 BC. Dating by carbon isotope analysis is relatively easy from charcoal in the slag. The earliest evidence comes from Afghanistan, where tin is found with copper ores, and bronze producers in the Near East seem to have been supplied with tin from Soviet Central Asia. The early bronze age in Cornwall and Brittany (about 2000 to 1500 BC) has bronze objects with a high tin content, indicating a local abundance of tin. In China, where tin was also plentiful, sites have been confirmed at 2000 BC. The bronze age began in South America two to three thousand years later but wrought bronze objects are typical of the early bronze age elsewhere. Low tin alloys have been found in nearly all early civilisations but only as trade improved between tin and copper pro-

LEFT: *Bronze Inca axe in the Camborne School of Mines Museum. Found by Mr Lloyd Rundle near Zaruma Mine, Ecuador, in 1920 it was sent to Cornwall from the USA by the Rundle family. They had traced their family tree to a miner who had left Cornwall for America towards the end of the eighteenth century. At that time Cornish mines were closing and there was an exodus of 'Cornish Jacks' to other mining regions throughout the world.*

RIGHT: *A Brunton calciner used to treat impure ores. Remains of these can still be seen in Cornwall, for instance at Wheal Peevor near Redruth, but are usually overgrown and difficult to distinguish.*

ducing areas was the standard 7 to 10 per cent tin bronze achieved. Throughout the period there is a marked increase in the quantity and quality of metal objects for daily as well as ceremonial use.

In Europe the Greeks and Romans developed bronze for use in architecture and engineering. Pliny writes of tin-lead solder and the use of tin coating on copper. In the fourteenth and fifteenth centuries tin plating developed in Bohemia on hammered sheets of iron and tin plating in Britain was granted a royal charter in 1670. During the seventeenth and eighteenth centuries European tin works were gradually absorbed by the English tinplate industry. In 1800 there were eleven works in south-west Britain using cheap wrought iron and Cornish tin although by 1880 most of the tin was imported and the local industry began to decline. The tinplate industry in the United States did not become successful until after 1890 but grew rapidly in the early 1900s.

The main centres of tin production in Europe were Devon and Cornwall, the Iberian peninsula, Bohemia and Saxony. The peak of production in the Bohemia-Saxony region was in the sixteenth century but the decline caused by the Thirty Years War left Devon and Cornwall as the principal western producers. By this time lode tin was being mined, and therefore grinding and washing processes were required as well as smelting. In Cornwall most of the early workings have been obliterated by later mining but in Devon and Bohemia these may still be seen. There are remains of blast furnaces and granite moulds for casting ingots very similar to those described by Agricola in 1556.

Cornish smelters or *blowing houses* (huts made of stones and turf) regularly caught fire, either by accident or deliberately to recover the tin dust. The furnaces were made of large stones held together with iron clamps and had bellows driven by waterwheels. The metal was ladled into iron pots for refining and then cast in granite moulds. While very pure alluvial ore was being used ingots of 99.9 per cent tin could be produced with these primitive charcoal furnaces. The process was so efficient that early slags often have less tin left in them than modern ones.

Alluvial tin ores used little fuel because only one process was needed — reduction from the oxide — but mixed ores required roasting to remove impurities and used correspondingly more fuel. Before 1700 the ore was dry-stamped to a fine sand, then added directly to the fuel

RIGHT: *A smelter from Saxony wearing the typical dress of the period. The leather apron and hat provide protection from the heat. The miners of Saxony were very proud of their profession and had symbolic uniforms for wearing in parades, each part of the mining operation having a different dress.*

BELOW: *The Chyandour tin smelting works near Penzance, Cornwall, in the late nineteenth century. Tall chimneys were needed for the up-draughts but here they are not as substantial as the stone chimneys of the engine houses. The concentrate can be seen stockpiled and a workman wearing an apron is leaning against the doorway.*

(wood, peat or charcoal) in the blowing houses. By 1703 coal was being used but this contaminated the tin and more elaborate smelting procedures had to be devised. In the mid eighteenth century pumping with new steam engines enabled lode tin to be mined from areas prone to flooding.

The increased amount of impurities led to complaints from London pewterers that the quality of the metal was poor. Calciners and roasting techniques had to be developed to make a product comparable in purity to that of alluvial ores. The remains of Brunton calciners, used to treat impure ores, can still be seen in Cornwall.

The main development of the industrial revolution was the gradual application of the reverberatory furnace to smelting. In the early nineteenth century the number of smelters increased in Britain as tin ore began to be imported. Then followed a decline as the tin producers of Malaysia built their own smelters. Gradually Cornish smelters closed, leaving only one in the whole of Britain. In the 1850s the United Kingdom supplied one third of the world's tin. By the 1880s it was overtaken by Australia and Malaysia. After 1900 both Indonesia and Bolivia greatly increased production, as did Nigeria and Zaire a little later. In China, where there are very large reserves of tin, production has fluctuated markedly since the late nineteenth century, but little is known of the industry in other eastern bloc countries. Before 1930 half the tin produced was handled by Malaysian smelters and a quarter by the United Kingdom but during the Second World War 40 per cent was diverted to the United States. In the 1980s a few large smelters handle most of the ore.

LEFT: *Tin ingots on the quay at Penzance, late nineteenth century. Cornish tin was shipped, in the form of ingots, all over the world. As Cornish smelters closed down the concentrate was sent by sea or rail to smelters in other parts of Britain. Today the concentrate is transported by lorry to the smelter in Humberside.*

BELOW: *Cornish smelting-house marks on ingots. The smelters produced slightly different grades of tin and each had its identifying mark. The lamb and flag symbol was commonly used though a pelican symbol was sometimes substituted, reputedly for trading with Muslim countries.*

The ruins of the mill at Poldice Mine, near Carharrack, Cornwall. This is now a desolate wasteland where once hundreds of miners worked both above and below ground.

THE CORNISH MINER

The image in the minds of most people of the work of the nineteenth-century Cornish miner has been influenced by picturesque ruins, curious names which hint at riches — Wheal Metal, Wheal Prosper and Wheal Plenty, and by romantic historical novels. In reality it was hard, dangerous and disease-ridden and brought few rewards for the miner or his family.

During most of the nineteenth century a Cornish miner's working year was divided into periods of one or two months at the end of which both his pay and his prospect of further work were not certain. Each working period began with a *setting* or *setting day*. For some time before the setting the mine captain or agent would go through the mine and decide the work to be carried out during the coming period. He would determine whether new shafts needed to be sunk or levels driven to develop the mine and he would also judge how many *pitches* or divisions of the lodes were to be worked or *stoped* in order to produce the ore. He estimated the quantity and quality of ore each pitch was likely to yield and the number of miners necessary to complete the work. He also calculated how many men might be needed for development work throughout the mine.

On the setting day the men employed at the mine, together with many others hoping for work, would gather outside the count house or mine office. The captain, standing on a raised platform, described the work to be done in the forthcoming period. Development work or *tutwork* was a system of piecework and payment was based on the amount of driving done or rock removed. Mining the ore was carried out by *tributing*, where the miners earned a fixed percentage of the value of the ore they raised. The captain then declared that in the previous period a miner and his pare (a team consisting of up to eight men) had carried out a certain task for a given payment. He asked the miner if he was prepared to take on the work again for the same terms and then inquired if anyone was prepared to work for less. There followed a kind of auction in which the bidding proceeded downwards and the work was 'knocked down' to the

11

Three engine houses of Wheal Peevor near Redruth. On the left was the winding engine which raised and lowered the skips, in the centre is the large pumping engine house which was always working to keep the lowest levels of the mine dry, and in the distance, further away from the shafts, is the stamps engine house.

lowest bidder. At the end of the period miners were charged for all the materials they had used, such as tools, candles and powder, and for the cost of hauling waste rock to the surface. They might also be debited with a subscription for medical attention and for the maintenance of a club to provide help for their families should they fall ill or suffer injury. At best the tribute system made the miner's income uncertain; at worst a bad period might end with the miner owing money to the owners for his tools and supplies.

If the miner and his pare were fortunate and won their contract the team would be divided, usually to work in three eight-hour shifts. Because the tribute system encouraged men to move from mine to mine to find employment or improve their pay a walk of 8 or 10 miles (13-16 km) was not uncommon before the shift started. Then began the awful descent on wooden ladders — 1000 or even 2000 feet (300-600 m) — and the long walk to the dark and foul-smelling pitch where the work was done. For safety the miner had only a felt hat, moulded to his head, and a pair of heavy boots. His light came from smoky tallow candles, one fixed to his hat with a lump of clay, others on ledges in the nearby rock. Until the 1870s drilling the shot holes was done by hand. In single-handed drilling, much used in the narrow workings of the Land's End peninsula, the drill steel was held in one hand and struck with a hammer held in the other. In multi-handed boring, favoured in the Camborne mines, one man held the steel while two others beat the drill alternately. Accidents were commonplace.

Until the 1860s gunpowder was used for blasting. The holes were charged and sealed with clay stemming. The miners pushed a hollow quill or straw through the stemming and filled it with finely ground gunpowder. Then a candle was fixed under the protruding end of the straw in a way that would allow it to burn for a short time before the 'fuse' ignited. Premature blasts and misfires were all too frequent. The invention of the safety fuse by William Bickford in 1831 made blasting much less hazardous but gunpowder is a low-power explosive and it could only be used to break off large pieces of rock close to a free face. The miner and his team would have spent many hours breaking these up into smaller blocks which could be raised to the

ABOVE, LEFT: *One of the many derelict engine houses of Cornwall, Killifreth, near St Day, which ceased operating in the 1920s. The chimneys or 'stacks' are all different but this one with its slender, tapering upper section is quite distinctive. The high stacks were required for strong up-draughts to produce the head of steam and the top of this one seems to have been added later when even larger steam engines were introduced.*

ABOVE, RIGHT: *A group of miners with, probably, the mine manager. The canvas jacket and trousers, heavy boots and felt hats are considered the traditional dress of the Cornish miner. A sticky clay was kneaded into a ball to attach the tallow candles to the hats. Spare candles were carried round the neck or tied to the jacket where they were easily accessible.*

BELOW: *A group of miners at 'croust', eating Cornish pasties, the traditional food of the miner. Some men in the photograph have removed their hats and the skull caps worn underneath, providing extra protection and a better fit, are visible.*

ABOVE: *Cathedral Stope at the 400 fathom (730 m) level at South Condurrow mine near Camborne, photographed in 1892. This vast space has been mined in the granite without the need for props and is directly below the King Edward Mine buildings.*

BELOW: *In contrast to the photograph above, here in the 66 fathom (120 m) level at Blue Hills mine, St Agnes, the hanging wall in 'killas' (slate) is kept in place by pit props.*

The 412 fathom (753 m) level in Dolcoath, 1893. Two or three days after this photograph was taken, on 20th September, the roof collapsed killing seven men and trapping one miner for thirty-seven hours before he was rescued. At this point the lode was 30 feet (9 m) wide and dipping 48 degrees south. The timbers used were 20 inch (508 mm) square pitch pine, 34 feet (10.3 m) long and set at 3 foot (0.9 m) centres.

tramming level and then transported to the shaft. Furthermore, because the power of a heavy hammer blow is delivered on the downstroke, the miners had to work by underhand stoping, cutting the rock downwards from one level to the next below and continually having to clear ore and raise it from their workplace.

Two important inventions changed the technique of Cornish mining dramatically. The first, in the 1860s, was Nobel's development of powerful high explosives based on the dangerous and unstable liquid nitroglycerine. Weight for weight these were about three times as powerful as gunpowder. Fewer shot holes were needed and the violence of the explosion shattered the rock so that it was no longer necessary for the miner to blast close to a free face. The fumes were less noxious than those from gunpowder though the air was still foul in an unventilated pitch after shot firing and the miner and his pare could ill afford to wait until it had cleared.

The second development was the invention of the rock drill, driven by compressed air, in the 1870s. This eased the miner's burden greatly and it also changed the method of mining. The mechanical drill could be braced against the floor and work upwards into the roof or 'back' of the drive. Thus the technique of overhand stoping was developed, whereby the miners drilled and blasted the lode above their heads, working from the pile of broken ore which could be periodically drawn off from below and carried to the haulage shaft. In the early days mechanical drills were run dry, creating thick and dangerous dust, so that the miners traded one hazard for another more insidious one, which slowly killed them. Eventually, however, water was fed through the centre of the drill steel and this not only improved cutting but also laid the dust.

Even though the work was hard and dangerous perhaps the most debilitating aspect of the nineteenth-century miner's life was the appalling climb back to the

ABOVE: *Setting the fuse at South Condurrow, 1892. Candles seem to have been used in Cornwall until the 1910s or 1920s, when carbide lamps were introduced.*

BELOW: *Climbing to grass at South Condurrow. In the background is the inclined skip road.*

Multi-handed boring at Dolcoath Mine, near Camborne, at the 375 fathom (686 m) level in March 1904. The man in white examining the rock face would be the underground mine captain. The temporary platforms using broken ladders and planks were erected hastily by the miners without heed of safety as speed was essential.

17

surface at the end of an arduous shift. In winter he might have seen the daylight only on Sundays and occasional feast days. Terrible accidents and crippling lung disease were rife among the miners of the nineteenth century.

The first *man engine* was installed at Tresavean Mine, near Redruth, only in 1842. A man engine was an assembly of wooden rods, with hand and foot holds cut in it, which extended down the shaft. Platforms were fixed in the shaft every 10 or 12 feet (3.0 to 3.7 m). The rod assembly was connected to a Cornish engine with a stroke similar to the plat- form spacing so that the miners could be raised or lowered by stepping on and off the rods and on to the platforms on alternate strokes. The miner's lot was eased but even the man engine took its toll. At Levant Mine, St Just, in 1919 the beam collapsed, the riding miners were hurled down the shaft and thirty-one were killed and many others injured. It was only from about 1900 that the miners began to be afforded similar treatment to their valuable ore and could travel quick- ly in cages to and from the depths of the mines.

LEFT: *The Dolcoath 'man engine', by which a miner could reach the surface by stepping on to the moving rod on its up-stroke and off again on to a platform as the rod went down. At the beginning of a shift it was probably quicker to use the ladders to reach the working end, as the two miners in the photograph are doing.*

RIGHT: *'Skip-riding' was an illegal practice as there was no overhead protection. Later iron 'coffins' with a roof over were provided. This photograph was taken in Dolcoath mine in the late nineteenth century. Unlike in coal mines, smoking was permitted in Cornwall as there was no danger of explosions from methane gas.*

A tin dredge operating in Malaysia, 1980. Output depends simply on the size and speed of the buckets.

TIN MINING

Today tin ores are mined by three different methods: from veins and lodes by hard rock underground mining; from alluvial deposits by hydraulicking and dredging; and from shallow seabed deposits by dredging. Most of the world's tin comes from poorly consolidated alluvial occurrences. A dredge is in effect a floating mine which moves across the surface of an artificial lake using a bucket chain to excavate the tin-bearing ground. The ore is carried up and on to the dredge, where the cassiterite is separated from the waste. The first successful dredges were used in Thailand in 1907 and since then they have become ever larger and more powerful. The largest in use are about 200 feet (60 m) long. An increasing proportion of tin is mined by offshore dredges in calm near-shore waters of the continental shelf.

The most important method of mining tin in south-east Asia is by hydraulicking. Very powerful jets of water are sprayed from monitors with such force that the tin-bearing alluvium is broken up and can be transported by a gravel-pumping system. The heavy cassiterite grains are separated under gravity from the lighter waste material. Some ore is obtained by employing a natural head of water rather than power-driven jets.

Modern underground mines are usually deep because most near-surface deposits were exhausted long ago. About one fifth of the western world's supply of tin comes from underground and most of this is from Bolivia and Tasmania. The mines may be of any size, ranging from highly complex workings at great depths with many miles of tunnels to small simple operations employing just a few men.

In Cornwall tin mining began with tin streamers searching the riverbeds for dark heavy pebbles of cassiterite. Sometimes they may have selected likely areas to dig small pits, which they lined with timber to prevent them from collapsing. If tin was found more pits were dug to determine the extent of the ore-bearing ground. The tin ore was taken out, then washed down to a channelway or *leat* at the lowest end of the working, where it could be concentrated and collected. If there was too much water a simple pump, driven by waterwheel or windmill, may have been used to keep the ore uncovered.

The ability to work these deposits

19

ABOVE: *South Crofty mine near Camborne opened in the mid nineteenth century. The stockpiles of ore on the skyline (right) are waiting for treatment in the mill (left). The shaft head-gear stands above the mill in the background. Old dressing floors can be seen in the foreground.*

BELOW: *Tin streamers at Redmoor, Lanlivery, taken between 1900 and 1914. On the right miners with picks are digging out the alluvial tin ore which is wheeled away in barrows to be further concentrated.*

ABOVE: *Ancient stone querns for grinding tin found at Vorvas, Lelant, in 1907. The small rounded stones seem to have been used for grinding by hand while the large querns may have been turned by a pole. Granite was used for this work as it is hard and occurs as blocks and rounded boulders.*

BELOW: *Tin washing 'buddles' at St Agnes, Cornwall. The finely ground ore was fed on to the sloping sides and water was run down the leat to wash the lighter gangue minerals away. The centre of the buddle rotated and the heavy cassiterite grains which remained were collected by a stationary brush and fed into a separate channel. This operation needed little labour and was very efficient. The remains of stone buddles can be seen on the cliff tops at Botallack but most of the wooden structures have disappeared.*

ABOVE: *A horse whim in use at Wheal Busy, Chacewater, Cornwall, in 1907. Ore is brought up in a 'kibble', or iron bucket, attached to a chain around the barrel, which the horse turns. In the background a second kibble is emptied into a truck, which runs on rails to the left of the picture.*

LEFT: *Saxon miners cutting pit props in the Freiberg metal mines, 1891. The style of dress is different from the Cornish with long boots, leather aprons and conical hats. The walls are supported by brickwork.*

depended on the tools available. The Cornish miners used simple wooden shovels and picks, which were later clad with metal. These were adequate for working weathered near-surface lodes. Sometimes such 'mines' may have been 50 feet (15 m) deep. As tools and methods improved and explosives were introduced, harder and deeper rocks could be tackled. With the start of deep mining large quantities of waste were thrown out into rivers and streams. This in turn led to a new industry because the waste could be re-treated by tin streamers working downstream of the mines to extract any tin which may have escaped. The problem for the miners was that the ore had to be crushed to separate it from the gangue, but the finer the ore was

22

ABOVE: Miners emptying ore from a kibble, or iron bucket, which had three rings, two on top and one below, attached to chains. As the lower chain was pulled the kibble tipped over. The trucks ran on rails and were usually pulled by the miners although horses were sometimes used.

BELOW: Young miners at work. Boys started work early, carrying tools for their fathers and older brothers. As soon as they were strong enough they were proud to become miners like the rest of the family.

ABOVE: *A large stope in a Cornish tin mine. The magnesium flare used to take the photograph gives a false impression of conditions underground; pinpricks of light in the background are candles which barely threw any light.*
BELOW: *A raise-borer in a modern tin mine at Wheal Jane, near Chacewater, Cornwall, 1984. This machinery has taken the place of the old miners with hammers and picks.*

crushed the more was lost in the waste. The tin streamers became skilled in extracting the fine tin which the miners did not or could not recover.

Primary tin deposits are usually found in steeply inclined lodes. In modern mines these are sometimes worked to depths of several thousand feet below the surface. In a typical mine shafts are sunk to one side of the lode. Tunnels are then driven at regular intervals from the shaft to meet the lode. From these further sets of tunnels are driven along the ore body. The miners drill upwards into the lode, charge the drill holes with explosive and blast the rock so that it falls and makes a platform from which to work. In this way mining advances from one tunnel or *level* to the one above. Ore is drawn off at the bottom of the working or *stope* and transported to the shaft in small trucks usually hauled by powerful battery-driven locomotives. Sometimes the rock is sufficiently strong to allow the lode to be extracted without the need for support. Elsewhere, if the rock is weak, the miners must leave pillars between the stopes to provide support. These pillars may contain rich ore and are often removed in the final stages of a mining operation.

Chinese women tin washers working tailings using wooden pans or 'dulangs' in Malaysia, 1950. This type of tin extraction has been carried out for hundreds, if not thousands, of years all over the world. Even now a Cornish miner may pick up a vanning shovel and separate out the tin in this manner as a quick method of assaying the ore.

PROCESSING AND SMELTING

Mineral processing or 'ore dressing' is the collective name given to a number of methods of separating valuable constituents (payable minerals) from the valueless (gangue) minerals in an ore. The processes that are used exploit differences in the physical and chemical characteristics of minerals — differences in density, solubility, magnetic properties and surface chemistry — to produce a *concentrate* of the payable minerals in a form suitable for recovering the metal. The gangue minerals form the *tailings*, which have to be stored in dams.

The processing of tin ores in the past was relatively simple since they were found either in alluvial deposits or near the surface of lode-type deposits. In both cases erosion by wind and water and oxidation had broken down the ore, removing some unwanted elements such as sulphur (normally associated with iron, copper, arsenic and zinc) but leaving the tin oxide unaffected. The ore was concentrated by washing it in pans, on a small scale, or in sluices on a larger scale. The methods depended on the large differences in density between cassiterite and its associated gangue, the former sinking to the bottom of the sluice or pan and the gangue being washed away. These methods can still be seen in use in many parts of the world such as Namibia, Malaysia and Brazil.

Little is known of ore dressing in Cornwall before the sixteenth century but we know that Queen Elizabeth I sent for miners from Germany to develop mines and smelters in the county. The machinery and methods they introduced are those illustrated by Agricola in 1556. In 1602 Carew described how the ore was broken by hammers before being further crushed in Cornish stamps. These consisted of three to six logs mounted vertically, with the ends bound with iron. They were lifted up by projections on a wooden cylinder which was rotated by a waterwheel and dropped on to an iron plate on which the ore had been placed. The ore was ground finer between grinding stones turned by water, then diverted over a series of green turfs in which the heavy cassiterite grains were trapped.

In 1758 Borlase told how ore was sorted underground by hand and then brought to the surface for further sorting and breaking. This work was done by young boys and by girls known as 'bal maidens'. Mechanisation came with the industrial revolution. The steam engine gave rise to steam-driven stamps, continuous jigs and, later, crushers, shaking tables and vanners. These last two were used to separate fine particles.

Sulphide and tungsten minerals became a problem as the mines became deeper. Their specific gravities are close enough to that of cassiterite to make efficient separation very difficult. They were therefore concentrated with the tin oxide and this concentrate was roasted to drive off sulphur as sulphur dioxide and convert the iron to iron oxide, which was easily washed away. At first this was done on a simple open hearth but by 1820 continuous calciners had been developed and continued in use for about one hundred years. In the 1920s the flotation process (exploiting differences in surface characteristics) took over from roasting to remove sulphides.

In a modern mine the first stage of crushing takes place underground in a *jaw crusher*, and then the ore is trans-

ABOVE: *Sixteenth-century tin stamps illustrated by Agricola in his 'De Re Metallica', 1556. Ore was fed under the hammers, which were raised and dropped by rotating poles with projections. These were driven by waterwheels.*
BELOW: *'Bal maidens' in their best clothes at Dolcoath mine near Camborne in the late nineteenth century. These ladies usually wore coarse aprons and sacking over their heads to protect them from the dust. Their job was to hammer the ore to a suitable size for the stamps. They are seen holding the typical Cornish long-handled shovel.*

ABOVE: *Cornish stamps at Levant, late nineteenth century. These had improved since the sixteenth century with the introduction of cast iron for the stamps, cog wheels, and a much larger battery driven by the steam engine in the background.*

BELOW: *Frue vanners at South Crofty mine, Pool, Cornwall, in the late nineteenth century. Fine ore is fed on to the moving belts and as water passes over it the unwanted light material is washed away and the tin collected at one end of the belt.*

ported to the surface and crushed finer. It may be washed and screened to remove fine particles before being treated in a *heavy medium separator* (a bath filled with a suspension of very fine, heavy particles in water). The waste material floats and is removed. The sink product from this is ground in tumbling mills and then is sorted into a number of size fractions, which are treated on spirals and shaking tables. Each plant has a section for treating *middlings* (the particles containing both payable and gangue minerals), which have to be ground finer. The concentrates are then passed to flotation machines, where the sulphide minerals are floated off. Gravity concentration is not always effective so flotation is used to recover cassiterite particles finer than 30 microns (one micron is $\frac{1}{24,000}$ inch). Chemicals are added which make the surfaces of the cassiterite particles, but not the gangue, water-repellent. The unwanted minerals remain in the machine and are discharged to tailings dams, where the water is removed and recycled to the plant. When the dams are filled and have dried out they are grassed over and landscaped.

The processes involved in tin smelting are complex. The high temperatures needed to reduce tin oxide with carbon also reduce other metals and these form compounds with the tin. Smelting is done in two stages. Primary smelting takes place in a reverberatory furnace or a blast furnace. Secondary smelting involves the re-treatment of the slags to extract further metal. In a tin-smelting furnace heat is reflected from the roof so that fuel does not come into contact with the charge. The base slopes down to a tap hole, which can drain the furnace completely. The molten tin is cast into slabs and the slag overflows and is run off for re-treatment. After casting, the tin must be further refined by heat or electrolysis. The tin is reheated to just above its melting point and then run off, leaving behind other metals with higher melting points. Finally it is boiled with steam or compressed air or stirred with green wood poles; any impurities form a light scum, which can be removed. The tin is cast in ingots: grade A exceeds 99.8 per cent pure tin.

A shaking table at Wheal Jane. The dark streak across the table is the heavy cassiterite concentrate, which is collected by the chute in the foreground. The very light coloured gangue minerals are taken away for disposal in the tailings dam.

Pewter items from Truro Museum, Cornwall. A church flagon, c 1634, an eighteenth-century English chalice (one of a pair) and a bleeding bowl, c 1880.

USES OF TIN

As a metal, tin is invaluable because of its special properties. It is non-toxic, resistant to corrosion and has a low melting point. It is soft, withstands friction well and has a bright, attractive appearance. Important tin deposits, unlike those of many other metals, are confined to a small number of developing countries: one of the largest consumers, the United States, has no important tin deposits of its own. Although it is a material with a long history of use, the importance of tin has not diminished during the modern expansion of technology and science. Some tin-based products have disappeared, like the tin coins of south-east Asia, and others, such as pewter, have become much less common. However, the total range of products has increased greatly. The tinplate industry accounts for about 40 per cent and solder a further 20 per cent of its use.

Tinplate is made of mild steel with a very thin coating of tin. The process originally involved dipping iron plate into a vat of molten tin and draining off the surplus for re-use. It was not until much later that a great impetus to this industry occurred when long-term food preservation was invented in France. This stemmed from the realisation that contact with air led to food decay. Non-toxic containers were needed and the first tin-canned food was produced about 1912. Mechanisation of the tin-plating process followed but not until the 1940s was the process drastically altered by the introduction of electrolytic methods of tin-plating on to steel.

Over 90 per cent of tinplate is used for making cans, mainly for food and drink, while less is used for other products such as paint containers and aerosols. Since the 1950s there has been an enormous increase in the demand for tinplate as the food-canning market expanded. At the same time non-container uses have expanded to include electronic equipment,

29

reflectors, toys, batteries and electric fires. The increased use of tinplate worldwide has greatly improved the standard of living for many people.

Nearly a quarter of tin produced is used in the form of soft solder, an alloy of tin and lead with a low melting point. This is important in the electrical and electronics industries, but by far its oldest use is in plumbing.

Another alloy widely used in past times is bronze. Now bronze is mainly used artistically for such things as ornaments, musical instruments and kitchen utensils, though its great resistance to corrosion allows its use in marine bearings and pumps.

Pewter is a lead/tin mixture, usually about 20 per cent lead, although modern pewter for decoration may contain over 90 per cent tin. Britannia metal, a more recent equivalent, is an alloy of tin and antimony. Other alloys have been developed for a wide variety of uses: tin/nickel for electronic printed circuits, tin/zinc for hydraulic equipment and tin/aluminium for bearings. Bell-metal also contains tin, while the finest organ pipes are made of almost pure tin.

A modern development has been in the chemical industry, where organotin compounds have found use in plastic manufacture, in disinfectants, wood preservatives, fungicides, marine paint and stabilisers. The first organotin compound was discovered in the 1850s; however, it was not until the 1960s and 1970s that production began to expand. Inorganic tin compounds were used in Egypt for dyeing as long ago as AD 500. Tin, in the form of stannic chloride and stannic oxide, is now produced for the manufacture of catalysts, veterinary medicines, toothpaste, soap and colourants.

A newly developed use, discovered in the United Kingdom, has been in coating glass with tin oxide to add to its strength for plate glass. The addition of a small amount of tin to cast iron improves its hardness and ease of casting for use in the car industry. The low coefficient of friction of tin is exploited in high-speed machinery and it is widely used in alloys for making bearings.

Although new uses for tin have emerged, there has not been a substantially increased demand for the metal as these new processes require only small amounts of primary tin. While the market has grown, the share of the metal market has been reduced; for example, at one time all metal cans were made of tin but now many are made of aluminium. Tin tubes for medicinal purposes are still used, but in the printing industry changes in type face and computerisation have almost eliminated the need for tin.

Secondary or reclaimed tin metal comes from tinplate works and canning factories. Tin has long been recovered from this clean scrap but no longer in such large quantities as previously, when the plate was dipped rather than electrolytically coated. Of the tin used for plating only about 10 per cent is recycled, the rest being thrown away. About one third of the world production of tin is used once, then discarded on rubbish tips. In spite of years of research and money spent on finding ways to de-tin used cans, a successful and economically viable method has not yet been developed. However, alloys such as solder and bronze may be recycled many times without the need to separate the tin. Meanwhile conservationists attempt to encourage preservation of this important and non-renewable resource.

Until October 1985 an international agreement among tin producing and consuming nations operated to buy up all surplus production. This created an oversupply and allowed the price to rise to an artificially high level of over £10,000 per tonne. That October the agreement was suddenly terminated and the tin price crashed to under £4,000 per tonne. Many producers throughout the world were forced to close down. Underground mines with higher costs were especially affected. In Cornwall, prior to mid-1985, there were three producing underground mines, three developing mines and a variety of surface mining and dredging operations. Only the three producing mines survived, but had to reduce their labour forces. Wheal Jane and South Crofty Mines received susbstantial government aid and have pursued a programme of modernization to reduce costs. That programme appears to be working successfully.

The tailings dam at Wheal Jane, 1984. The fine waste material settles out of suspension behind the dam and the water is recycled to the mill.

FURTHER READING

Agricola, Georgius. *De Re Metallica*. Translated by H. C. and L. H. Hoover. (First published by *The Mining Magazine*, 1912.) Dover Publications, 1950. (A sixteenth-century account of mining techniques.)

Atkinson, R. L. *Copper and Coppermining*. Shire Publications, 1987.

Barton, D. B. *A History of Tin Mining and Smelting in Cornwall*. D. Bradford Barton Ltd, 1967. (Description of Cornish mining.)

Barton, D. B. (editor). *Historic Cornish Mining Scenes Underground*. D. Bradford Barton Ltd, 1967. (Burrow's nineteenth-century underground scenes.)

Crowley, T. E. *Beam Engines*. Shire Publications, 1976. (Description and history of the beam engine.)

Earl, Bryan. *Cornish Mining*. D. Bradford Barton Ltd, 1968. (Techniques of tin and copper mining.)

Hedges, E. S. (editor). *Tin and Its Alloys*. Edward Arnold, 1960. (Description of alloys and their uses.)

Mantell, C. I. *Tin: Its Mining, Production, Technology and Application*. Hafner, Publishing Company, 1970. (Comprehensive cover of the tin industry.)

Penhallurick, R. D. *Tin in Antiquity*. The Metals Society, London, 1986. (Worldwide coverage of the history of tin in ancient times.)

Robertson, William. *Tin: Its Production and Marketing*. Croom Helm Ltd, 1982. (Study of world tin markets.)

Trounson, J. *Mining in Cornwall*. Volumes I and II. Moorland Publishing Company, 1980. (Photographs of early mining scenes.)

Tylecote, R. F. *A History of Metallurgy*. The Metals Society, London, 1976. (Tin workings from earliest times described.)

Willies, Lynn. *Lead and Leadmining*. Shire Publications, 1982. (Description of steam engine and mine plan.)

MINING SOCIETIES

Carn Brea Mining Society: c/o Camborne School of Mines, Pool, Redruth, Cornwall TR15 3SE. Telephone: Camborne (0209) 714866.

Plymouth Mineral and Mining Club: 25 Budshead Road, St Budeaux, Plymouth. Telephone: Plymouth (0752) 361375.

Royal Geological Society of Cornwall: West Wing, St John's Building, Alverton Street, Penzance, Cornwall TR18 2QR.

Trevithick Society: Secretary, Bill Newby, 'Gonew Viscoe', Lelant Downs, Hayle, Cornwall TR27 6NH. Telephone: Penzance (0736) 740337.

PLACES TO VISIT

In the United Kingdom the following museums and areas are worth visiting. Although many of the old shafts have been filled or covered it is essential to take special care when visiting old mine workings as they can be dangerous. Never go underground alone or without a guide and the proper equipment.

MINING AREAS

Carn Brea area for old mine workings: many derelict engine houses and associated buildings.

St Agnes district, especially the coastal section from Chapel Porth to Perranporth.

Botallack to Cape Cornwall, for spectacular views of engine houses along the cliffs.

St Day area: famous old tin and copper mining district now undergoing extensive reclamation for use as an amenity area.

MUSEUMS

Intending visitors are advised to find out the times of opening before making a special journey.

Camborne School of M 1es Geological Museum, Pool, Redruth, Cornwall TR15 3SE. Telephone: Camborne (0209) 714866.

Cornish Engines, East Pool Mine, Pool, Redruth, Cornwall (National Trust).

Cornwall County Museum, River Street, Truro, Cornwall TR1 2SJ. Telephone: Truro (0872) 2205.

Geevor Mine and Museum, Pendeen, Penzance, Cornwall TR19 7EW. Telephone: Penzance (0736) 788030.

Morwellham Quay Open Air Museum, Morwellham, Tavistock, Devon PL19 8JL. Telephone: Tavistock (0822) 832766.

Poldark Mine, Wendron, Helston, Cornwall. Telephone: Helston (0326) 573173.

Wheal Martyn Museum, Carthew, St Austell, Cornwall PL26 8XG. Telephone: St Austell (0726) 850362.